Pomapoo 20 Selfie Milestone Challenges

Pomapoo Milestones for Memorable Moments, Socialisation, Indoor & Outdoor Fun, Training

Volume 3

AF875387

Todays Doggy

Dedicated To All of You Wonderful Owners and Fans

Introduction

Welcome to the Original Doggy Milestone Series™ where you are encouraged to create those special moments with your pup. We have composed the milestones in a way that challenges you to set the stage before taking your photos.

Use props and make it fun - be creative in setting up your photos. Get family and friends involved - take it out with you - use it in different places and settings - have a play with it and most importantly, have a good time!

Good luck and enjoy your photo fun.

I'm So Excited

and I Just Can't Hide It

Watching My Favorite Show

Talk Later

Stairs?

Hmmm

Is There a Lift Nearby?

If You Could Take Me For a Walk

That Would Be Great

Too Tired To Make It

Onto
The
Bed

Do You Like

My New Hair Style?

Raining ey?

I Think I'll Just Pee Indoors Today

My Autumn Turd

Now You See It

Now You Don't

Let's
Pretend
Nobody
Saw This
Mess

and Walk Away Quietly

Everything

I Do

I Do It For You

Staring Out The Window

Wondering What To Bury Next

Please Don't Ask Me

To Do Anything Right Now

I Like To Kiss My Reflection

Simply
Because
I'm The
Best

Finished a Whole Pizza.

DO I...

1) Stop
& Feel
Disgusted?
OR
2) Finish
Strong
With an Ice
Cream?

You Can't Mess With Me

I’m the Big Bad Boss

It's That Time Again...

Belly Rubsss

I'm Gonna Tell You This Once:

NEVER PET ANOTHER DOG

I'm Getting Used To

Being Pampered So Much

Your Bed

Is Now My Bed

I Walk Around On My Back Feet

Pretending To Be Human

Lightning Source UK Ltd.
Milton Keynes UK
UKHW020726241019
352156UK00006B/238/P

9 781395 633851